TEST TUBE BABIES

THE SCIENCE OF IN VITRO FERTILIZATION

• • • •

WRITTEN BY

TAMRA B. ORR

**BLACKBIRCH®
PRESS**

THOMSON
⎯ ✦ ⎯ ™
GALE

San Diego • Detroit • New York • San Francisco • Cleveland • New Haven, Conn. • Waterville, Maine • London • Munich

THOMSON
GALE

For more information, contact
The Gale Group, Inc.
27500 Drake Rd.
Farmington Hills, MI 48331-3535
Or you can visit our Internet site at http://www.gale.com

Photo credits: cover, pages 7, 8, 9, 10, 11, 13, 17, 18, 21, 22, 24, 25, 27, 31, 34, 35, 36, 37, 39, 40, 41, 43 © CORBIS; pages 4, 5, 12, 19, 32, 38 © PhotoDisc; page 14 © AP Wide World; page 15 (bottom) © SIU / PhotoResearchers, Inc.; page 15 (top) © VEM / PhotoResearchers, Inc.; page 20 © John Giannicchi / PhotoResearchers, Inc.; page 23 © Cortier / PhotoResearchers, Inc.; page 30 © Hank Morgan / PhotoResearchers, Inc.; page 28 © Blackbirch Archives

LIBRARY OF CONGRESS CATALOGING-IN-PUBLICATION DATA

Orr, Tamra.
  Test tube baby / by Tamra B. Orr.
    p. cm. — (Science on the edge series)
Summary: Examines the causes of infertility, the history of in vitro fertilization, the steps involved in creating a "test tube baby," and ethical questions the technology has raised.
Includes bibliographical references.
  ISBN 1-56711-788-0 (hbk. : alk. paper)
  1. Fertilization in vitro, Human—Juvenile literature. [1. Test tube babies. 2. Infertility. 3. Reproduction.] I. Title. II. Series.
  RG135 .O77 2003
  618.1′78059—dc21   2002011928

Printed in China
10 9 8 7 6 5 4 3 2 1

618.178
ORR
2003

THE SCIENCE OF IN VITRO FERTILIZATION

One of the strongest drives that human beings have is the natural desire to reproduce. To have children is a goal for many couples, who see it as a way to send part of themselves into the future. Adults often feel that children create the concept of family; they see their sons and daughters as sources of joy, surprises, and challenges.

Although people's eagerness to have a family has remained the same over time, life has changed a lot in recent generations. Stress levels are higher; exposure to chemicals and other toxic elements has increased. Less-than-healthful diets and a lack of physical

Many people want to have children.

Stress, caused by things like traffic, can affect a person's ability to have a child.

exercise have all contributed to lower reproductive rates, which means that more people are unable to have children. For a growing number of couples over the last several decades, it has been quite difficult or even impossible to have a child. These couples struggle with infertility. They are unable to conceive or carry a baby to term after they have tried to have a child for a year or more. Infertility can be emotionally exhausting for both men and women, who may feel disappointed, depressed, or angry.

Although the problem of infertility once had no real solution, there are now several possible ways to overcome it. Researchers and doctors from many different fields have worked to find help for desperate couples, and in doing so, they have made some amazing scientific advancements. ART, or assisted reproductive technology, has come a long way in the last 50 years or so—and many thousands of babies have been born worldwide because of it. Many infertile couples today rarely lose hope because new theories and techniques are developed all the time.

CHAPTER 1

IN THE BEGINNING

The quest to have a child has left many infertile couples in search of answers and has propelled scientists and physicians to work to find those solutions. In vitro fertilization (IVF) is one of the best methods available today to help infertile couples have a child. In natural reproduction, the father's sperm enters the mother's body during sexual intercourse. When the sperm joins with an egg that comes from the woman's ovaries, the egg is fertilized. IVF reaches this goal in a different way. The words in vitro mean "in artificial circumstances." Fertilization refers to the joining of egg and sperm. Put together, these two terms describe IVF—the ability to start the development of a baby outside of the human body and then implant it, or put it in, a woman's uterus to grow fully.

Although children born through the IVF procedure are sometimes referred to as "test tube babies," the name is not accurate. The phrase calls to mind an image of a baby that literally grows inside a glass. This is far from the case. In fact, a test tube is not even used. Instead, the egg is fertilized by sperm in a petri dish filled with nutrients. In addition, children conceived through the IVF procedure develop inside of their mother's uterus—not inside a piece of scientific equipment. In IVF, the eggs that eventually create a baby are simply fertilized in another place before they are transferred inside the woman's body.

For many years before IVF was developed, the only option for infertile couples who wanted to have children was adoption. Today, however, the choices have expanded far beyond that, thanks to the dedication and determination of several physicians who kept searching for new possibilities.

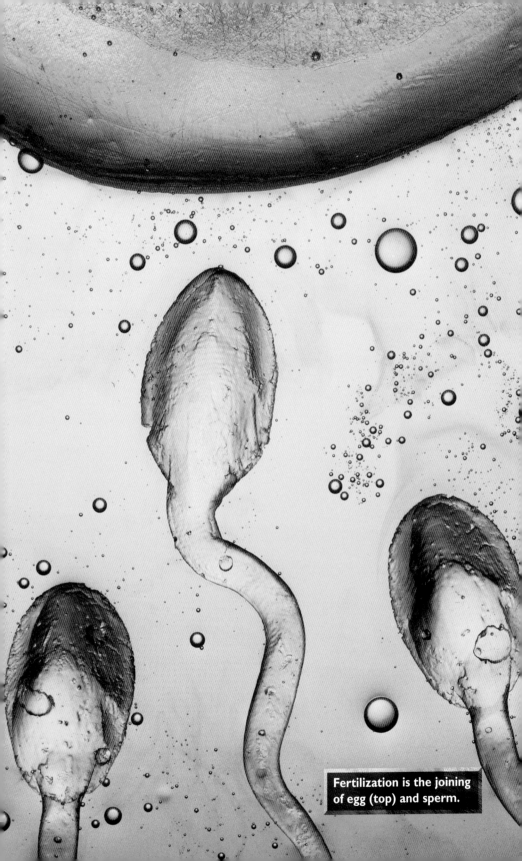

Fertilization is the joining of egg (top) and sperm.

EARLY EXPERIMENTS

One of those physicians was Dr. M.C. Chang, who began to experiment with rabbits in 1959. He hoped to find out if there was a way to create a rabbit embryo (a fertilized egg under eight weeks old) outside of the female rabbit's body. He took an egg from the female and put it in a petri dish with a male rabbit's sperm. To his delight, the egg and sperm merged to create an embryo. Chang replanted that embryo in the female rabbit's womb, and when a baby rabbit was born, Chang made history. His experiment was the first successful test of IVF ever recorded.

A technician injects sperm into a petri dish that contains an egg. In vitro fertilization (IVF) is the process of joining an egg and sperm outside of the body.

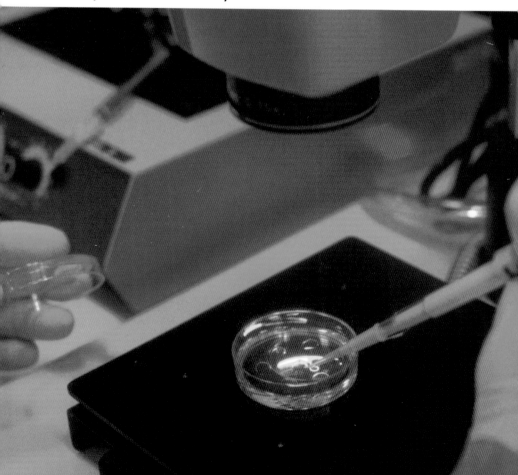

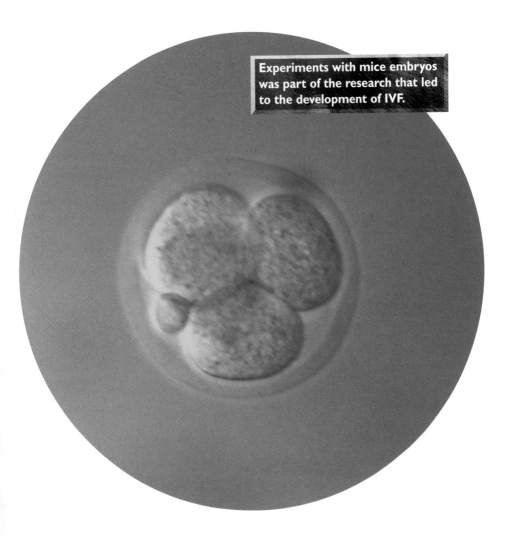

Chang's successes were watched closely by Dr. Robert Edwards in England. In the early 1950s, Edwards had already begun to think about whether it would be possible to use IVF on humans. He started experiments that involved fertilizing mice eggs in a test tube. Because the mice usually released eggs only during the night hours, Edwards often found himself in the lab at very inconvenient times. Three years of midnight visits led him to his next step, however. Along with his fiancée, Ruth Fowler, and his partner, Alan Gates, Edwards began to experiment with ways to control how many eggs the female mouse would release, as well as when she would release them. To do this, he used different concentrations of hormones. Finally, he found the perfect combination that would maximize the chance of pregnancy—and allow him to stay in bed at night.

Edwards carried out additional experiments on other animals, including rabbits, sheep, cows, and monkeys. Soon, his focus shifted to humans and what could be done to help them conceive. In order to test whether the idea and procedures he had used on animals would also work on humans, he needed eggs and sperm to be donated to his laboratory—something many people objected to. Some people believed that eggs and sperm are already a form of human life and should not be used in experiments in which they might be destroyed, thrown away, or misused. Edwards had other critics, too. Most gynecologists (doctors who specialize in women's reproductive health) scoffed at Edwards. They thought his ideas were preposterous.

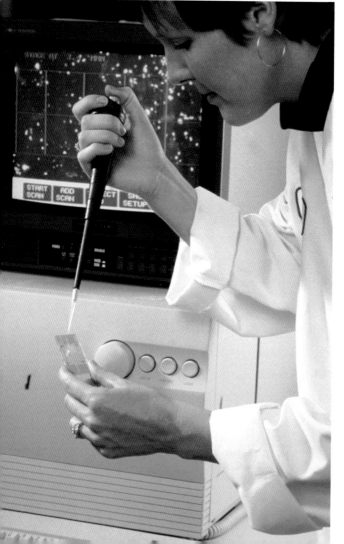

A scientist experiments with sperm. Some people believe sperm should not be used in experiments.

Edwards suffered through many disappointments before he found success. The attempt to find just the right balance of hormones resulted in mistakes and frustration. The entire process was one of trial and error. He would try something and see if it

Dr. Patrick Steptoe used a laparoscope to see inside a woman's reproductive system.

worked. If it did, he moved ahead one small step. If it did not, he had to start over again. Because he had very few eggs to work with, and did not know the many intricate ways that human eggs, sperm, and hormones might differ from those of animals, his experiments were both difficult and fragile.

In 1968, Edwards joined Dr. Patrick Steptoe, a world-recognized expert in laparoscopy—a surgical procedure that uses a laparoscope, or a slender, tubular instrument, to go through an incision to see inside the body. The laparoscope gave Edwards his first real opportunity to see inside a woman's reproductive system and remove eggs without doing any damage. With it, he was able to make even more steps forward. IVF was still a long way from becoming the procedure used today, though; it took years of experiments, for example, before the researchers could get the fertilized egg to grow past the first few cell divisions without dying. It took even longer to discover the best way to transfer the developing embryos safely into the woman's body.

Edwards and Steptoe worked as a team for the next twenty years. Together, they searched for ways to improve the IVF procedure. Each experiment taught them a little more about what to do—and not to do.

THE MANY CAUSES OF INFERTILITY

There are many causes of infertility. In 10 percent of the cases, the exact cause is never known. The rest of the cases, however, point to a wide variety of reasons.

Alcohol may cause infertility.

In men, the general reasons for infertility can range from a blockage of the sperm ducts to a history of sexually transmitted diseases (STDs). The use of drugs such as alcohol, nicotine, marijuana, and cocaine may also play

Nicotine can alter the genetic makeup of sperm.

a part, since they can sometimes alter the genetic material of sperm. Some men cannot reproduce because they have had a vasectomy that was not successfully reversed or because they have been exposed to toxic chemicals. Other men have sperm that is of low quality or insufficient quantity to conceive a child.

In a woman, there are even more things that can go wrong, so there are more potential causes of infertility. These may include scarring or blockage of the fallopian tubes due to STDs or surgery, structural damage from some kind of trauma, or uterine tumors. Endometriosis, a disorder in which the mucous membrane tissue grows outside of the uterus, can also make a woman infertile. Women who ovulate abnormally— too late, too early, or not at all—due to hormonal imbalances or genetics may also have difficulty conceiving. In recent years, many women have delayed pregnancy until after age thirty in order to pursue careers. It is often harder for women to become pregnant as they get older. In addition, drug use may affect fertility just as it does in men, since some drugs may alter the egg's genetic material. Other physical problems, such as an abnormally shaped uterus or an inability to manufacture the cervical mucus that ensures the passage of sperm, can also prevent conception. Some women may even face fertility problems because they are allergic to their partner's sperm.

In about one-third of infertile couples, there is a problem in both the man and the woman. In the other two-thirds of couples, the reasons for infertility are divided just about equally between men and women.

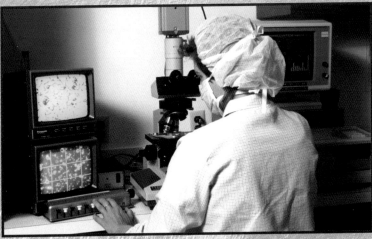

If scientists are able to diagnose the causes of infertility, they may be able to help couples have children.

THE FIRST TEST TUBE BABY

The lessons Edwards and Steptoe learned eventually resulted in an amazing breakthrough. The peak of their success came in July 1978 when Louise Joy Brown, a blond, blue-eyed baby who weighed 5 pounds and 12 ounces, was born in England. Her parents, Lesley and John Brown, were, like most new parents, very happy—but they had a special reason to rejoice. After Lesley had tried to get pregnant for years and had even endured a failed operation to repair her blocked fallopian tubes, the Browns had finally been able to have a child—something they had come to believe they could never do. Louise became the world's first test tube baby. Her birth caused an international sensation and caught the attention of infertile couples the world over.

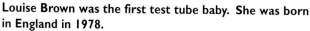

Louise Brown was the first test tube baby. She was born in England in 1978.

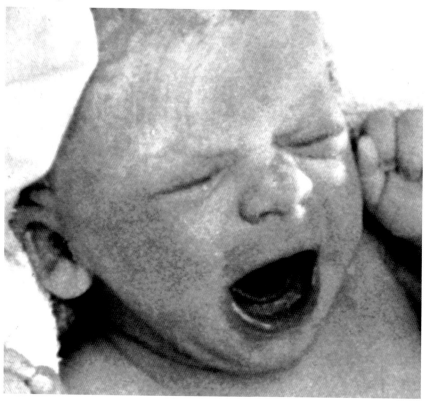

DEVELOPING USES FOR IVF

At first, IVF was primarily designed for women who had problems with their fallopian tubes. These tubes, which act as a highway through which the egg must travel to get from the ovary to the uterus, can sometimes become blocked, as was

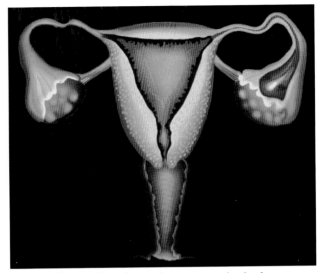

IVF was first designed to help women who had problems with their fallopian tubes.

true in the Louise Brown case. Often, the fallopian tubes are blocked by scar tissue from past pelvic infections, by a condition called endometriosis, or by excess fluid in the tubes, a condition called hydrosalpinx. When the tubes are blocked, an egg is not able to get through, which means sperm cannot get to it, and fertilization cannot occur. In IVF, there is no need to use the fallopian tubes.

Originally, a physician obtained the eggs that were to be fertilized through a surgical technique called laparoscopy. Physicians pushed a

A surgeon uses a laparoscope to obtain eggs from a female patient.

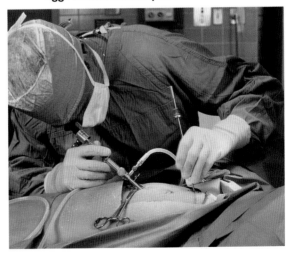

needle through the abdominal wall, then moved it through the bladder in search of the ovary. This was not only technically difficult, but also painful for the woman. Everything changed in the early 1980s, when doctors in Sweden discovered that they could use ultrasonic waves to get eggs, rather than cut into a woman's body. Doctors found that it was not only easier but also more effective to enter the woman's body through the vagina. This placed their equipment closer to the ovary, so the image was clearer. It also did not require the woman to be sedated.

By the early 1980s, doctors found that the IVF procedure could also help women who had hormonal or ovarian problems. In addition, IVF has been shown to help overcome male infertility issues. In a process called intracytoplasmic sperm injection (ISCI), a single sperm is captured and injected directly into the woman's egg. This way, even a man who has a low number of sperm or sperm of poor quality has a better chance to fertilize an egg.

In order to arrive at the relatively safe and highly effective method used today, IVF went through a strenuous process of experimentation. When it was first introduced, some people hailed the new IVF procedure as a miracle; others called it a curse. People worried that the process allowed doctors to interfere with nature rather than help it. Newspaper articles and letters reflected people's concerns about doctors who intended to play God. Many religions felt that it was a sin to create a human being in a laboratory, and believed the child would have no soul. The discovery and use of IVF raised many possibilities and controversies when it was first developed—and it continues to do so today. Despite the many disputes it has caused, though, there is no doubt that IVF has helped thousands of infertile couples have children through a relatively simple process.

Using IVF to create a baby has raised many ethical questions.

THE BIRTH OF ELIZABETH CARR

Roger and Judy Carr had been married for nine years. He was a mechanical engineer, and she was a teacher. They enjoyed a good marriage and had good jobs, and they felt that their lives were almost complete. Like many couples, however, they wanted a

In 1981, Judy and Roger Carr gave birth to the first test tube baby in the United States.

child. Judy had already had three miscarriages, though, and after the last one required surgery, she was told she would never be able to conceive.

At about the same time as Edwards and Steptoe were beginning to perform IVF successfully, a husband and wife team embarked on a new adventure. Gynecologist Howard Jones and his wife, endocrinologist Georgeanna Jones, had studied human fertility for thirty five years at Johns Hopkins University in Baltimore, Maryland. In the 1970s, they had decided to retire to Norfolk, Virginia, but their plans changed when they were invited to lead a group of scientists at Eastern Virginia Medical School. The topic of study was in vitro fertilization. They took the position, and it was not long before the Joneses met the Carrs, who had joined the group of couples willing to try the new fertility treatment that was making headlines.

The ultimate result of that meeting was Elizabeth Jordan Carr, who was born on December 28, 1981, and weighed 5 pounds, 12 ounces — exactly the same weight as her predecessor, Louise Brown of England. Just three years after the world's first test tube baby had been born in England, Elizabeth Carr became the United States's first test tube baby.

CHAPTER 2

THE CURRENT PICTURE

In vitro fertilization has opened new possibilities to childless couples throughout the world. Doctors have found a way to overcome the complicated problem of infertility through a fairly simple procedure.

Before the actual IVF process begins, a doctor does a number of tests to see if the exact cause of the couple's infertility can be pinpointed. It is common for a physician to order a semen analysis (report on the quality and quantity of sperm) for the man, a pelvic examination for the woman, and tests on both partners for sexually transmitted diseases (STDs) or signs of HIV, the virus that causes AIDS.

Many blood tests are necessary on both parents before the IVF process can begin.

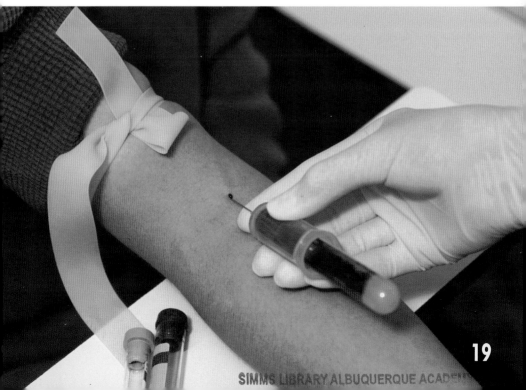

STEP ONE: PREPARATION OF THE EGGS

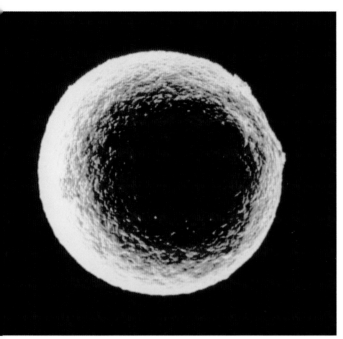

Drugs are used to trick a woman's body into releasing eggs.

The first step of the IVF process is to make sure that the woman's eggs are available and ready for fertilization. In the past, this meant that doctors had to wait to retrieve the eggs until a woman naturally released an egg as part of her monthly menstrual cycle. Today, however, they can administer hormonal medications instead. Often, these are painful daily injections that trick the woman's reproductive system and tell it to release far more eggs than usual. Taking hormones in this manner can be somewhat risky. The treatment can cause a condition called hyperstimulation, in which the body struggles too hard to handle the strain of releasing the extra eggs and to deal with the additional hormones that flood the system.

After the hormonal treatments, the couple watches for signs of ovulation (the release of eggs from the ovaries down through the fallopian tubes). This can be done with a special kit, or the woman's body can be monitored for a small rise in temperature. If a blood test confirms that ovulation is taking place, it is time to move on to step two of the IVF procedure.

STEP TWO: RETRIEVAL OF THE EGGS

In this step, an ultrasound probe in a sterile sleeve is inserted into the woman's vagina. When the follicles (small pouches that hold the eggs) appear on the screen, a needle is inserted alongside the probe. A local anesthetic is injected around the uterus, vagina, and cervix to numb the nerves in that area. The needle is then sent through the vaginal wall into the ovary, where it begins to pull out the eggs that are found there. Thanks to the hormonal medication that helps the woman's body produce extra eggs, it is not unusual for a doctor to retrieve up to thirty eggs—or even more—during this process. More than 90 percent of egg retrievals are now done through this ultrasound procedure. Because there is no surgery involved, it is faster and requires less recovery time than the older methods in which doctors had to cut into the body to get the eggs.

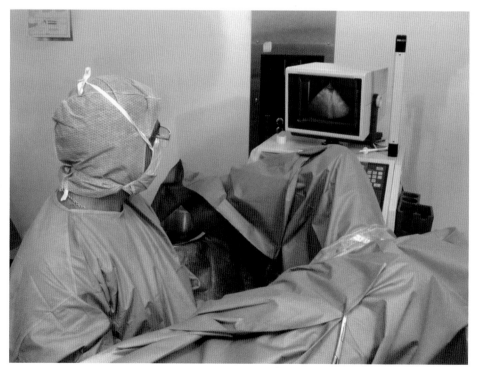

Retrieval of eggs from the woman is the second step in the IVF process.

STEP THREE: FERTILIZING THE EGGS AND CREATING AN EMBRYO

After the eggs have been removed from the woman's body, an embryologist—a doctor who specializes in the health of embryos—evaluates them. He or she determines how many eggs there are and how mature they are. The eggs are then placed in a petri dish full of liquid that bathes and nurtures them in the same way the woman's body would do.

While the eggs are kept warm in liquid, the male partner's sperm sample is prepared. First, the sample is washed in a special liquid designed to induce capacitation, the process that makes the sperm mature enough to penetrate and fertilize an egg. In nature, this process is done by a woman's cervical mucus; in a lab, this wash does the job.

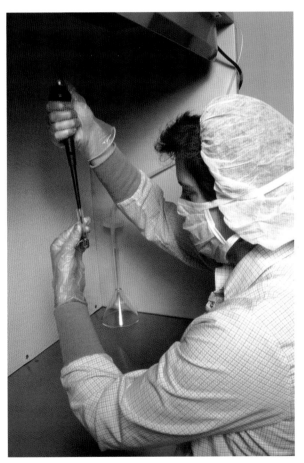

After the wash, the embryologist adds one to two drops of semen and puts the sperm in the dish with the eggs. Those drops usually contain about 50,000 to

An embryologist evaluates eggs to determine how many there are and how mature they are.

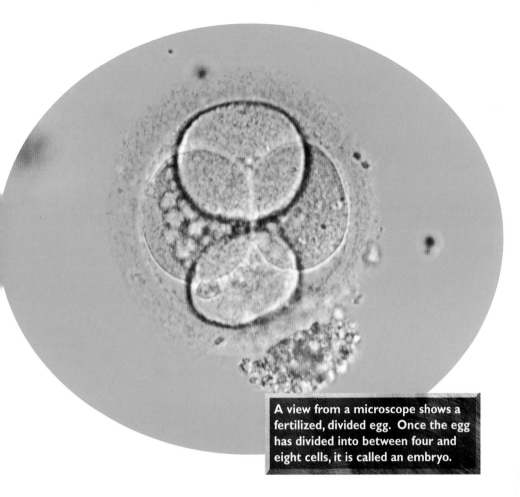

A view from a microscope shows a fertilized, divided egg. Once the egg has divided into between four and eight cells, it is called an embryo.

100,000 sperm. The dish is returned to the incubator until the next morning, when the embryologist takes a close look to see if any of the sperm have fertilized the eggs. If they have, there will be two new structures, called pronuclei (the singular is pronucleus) in the egg: one male and one female. Any cases of abnormal fertilization, such as eggs that have polyspermia, which means they have been fertilized by more than one sperm, are discarded. Unfertilized eggs may be frozen and saved for a later IVF attempt. Any fertilized eggs that are in good condition will be transferred to an environment designed to promote growth. Once a fertilized egg has divided into between four and eight cells—which takes place about 72 to 120 hours after the egg retrieval—the fertilized egg is called an embryo. It is then time for the vital fourth step.

STEP FOUR: TRANSFERRING THE EMBRYO

Finally, it is time for the newly formed embryo to be placed in the woman's uterus. This is an emotional time for many couples, since it calls on them to make a difficult decision about how many of the embryos they want to transfer. If ten eggs were fertilized, how many of those should be used? This question arises because doctors do not know how many—if any—of the embryos will successfully implant and result in pregnancy. A lot depends on the mother's age and condition. Often, the physician will help the couple make this delicate decision. For women under thirty five, most doctors recommend that only three embryos be transferred. It is commonly suggested that women between thirty five and forty have no more than four, while women over forty or those with repeated failed pregnancies are sometimes advised to transfer up to five.

A surgical team prepares to implant fertilized eggs in a patient. A successful implant will result in pregnancy.

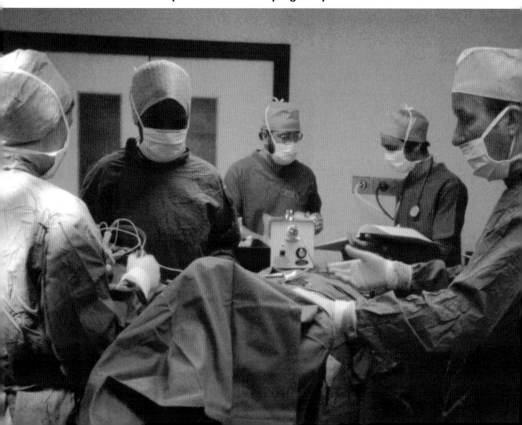

Twins are often the result of IVF pregnancies.

Approximately 20 percent of the time, at least one embryo will implant successfully. Besides the age of the woman and the condition of her uterus, the possibility of success also depends on the quality of the man's sperm, the history of each partner's drug use or STDs, and the quality of the laboratory and its staff.

The very real possibility that all of the embryos may implant brings up the concern of multiple pregnancies. England limits the number of embryos that can be transferred to three because of the risk that mothers will carry more than three at one time, which is dangerous not only to the mother but also to the babies. The United States and Canada, on the other hand, have no limit. Multiple births are common. One quarter of all successful IVF couples have twins; 5 percent have three or more babies. Multiple births are undoubtedly one of the most serious risks associated with IVF.

Once the number of embryos to implant has been decided, it is time to move ahead with the process. This portion of the procedure is often done in a dark, quiet atmosphere in order to help the woman relax. The woman's partner is usually allowed to be in the room with her as the incubator that holds the embryos is rolled in. An instrument called a speculum is inserted into the woman's vagina, and the embryos are drawn up into a small plastic catheter, or tube. The catheter is pushed up through the cervix into the woman's uterus, and the number of embryos agreed upon earlier is expelled. After it is withdrawn, the catheter is examined by microscope to make sure that all of the embryos have left the tube.

When the procedure is over, the woman lies down for one to two hours. This gives the embryo a better chance to implant without the downward pull of gravity that would occur if the woman was up and walking around. She is then sent home to rest for several days. After this, all the couple can do is hope and wait to see what a pregnancy test reveals in a couple of weeks.

WONDERS AND WORRIES

A couple must consider all of the risks of IVF and should discuss their concerns with their doctor.

The decision to have in vitro fertilization is not an easy one to make. Cost alone can be an inhibiting factor. When payments for an anesthesiologist, lab, physicians, and medications are all added together, IVF is a very expensive procedure. The cost often averages around $10,000, and few health insurance programs cover either IVF or any other fertility procedure. This means that the majority of the money must come straight from the couples who undergo the procedure. Besides cost, IVF demands a significant amount of time, dedication, and commitment from those involved. The four steps of the IVF process can be exhausting, both mentally and physically. In addition, a couple has to consider the risks associated with IVF, including the possibility that, even after all the effort and money are spent, the procedure may still fail to result in pregnancy.

Both an Australian study and one done by the U.S. Centers for Disease Control and Prevention (CDC) in March 2002 introduced yet another risk that couples must consider in regard to IVF.

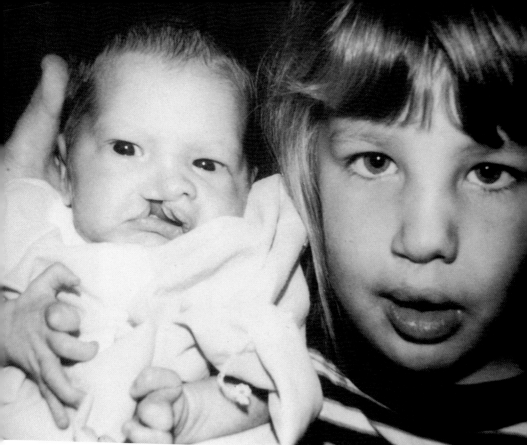

IVF babies may be born with disorders like a cleft palate.

According to the research, IVF babies tend to be smaller than babies born naturally, and they are twice as likely to have major birth defects diagnosed during the first year. IVF babies also tend to have multiple disorders, including cleft palates, clubfeet, and Down's syndrome. Scientists are not sure of the reasons for these findings. It is not clear whether the blame for the defects should be placed on the couple's original, underlying fertility problem; faulty lab procedures or medication; or complications that occur in the process by which the embryos are frozen and thawed. Researchers continue to look into the issue. In the meantime, the science and technology of IVF grows and improves all the time—bringing along with it more miracles and more questions.

FACTS AND FIGURES

- There are currently an estimated 6.1 million infertile couples in the United States.

- A quarter of the couples diagnosed with infertility have more than one factor that causes the problem.

- The average cost of the four-step IVF process is between $8,000 and $13,000. Very few insurance companies cover any portion of this cost.

- Approximately 5 percent of infertile couples attempt IVF.

- Seven percent of infertile women are Hispanic; 6.4 percent are white; 10.5 percent are African-American, and 13.6 percent are of other ethnic races.

- The average success rate for the IVF procedure is 22.8 percent. Some clinics have reported success rates as high as 40 percent, which is approximately double the success rate of attempts to get pregnant the natural way.

CHAPTER 3

A FUTURE OF POSSIBILITIES

In 2002, 50,000 babies were born worldwide through the IVF process. Clearly, IVF has enabled infertile couples all over the world to have children, but just like any other medical discovery, IVF has also opened the door to many other possibilities. Most people feel that some of these possibilities bring great hope for the future; others worry that the technology could also lead to profound dilemmas.

IMPROVEMENTS AND QUESTIONS

The IVF procedure has continued to improve since its initial use several decades ago. The trial-and-error battle to learn what hormones to use and when, how best to remove and replace eggs, and other such questions has been largely won. Still, researchers and physicians continue to work to improve the procedure. They focus in particular on how to make IVF safer and more successful.

Researchers continue to work to improve the IVF process.

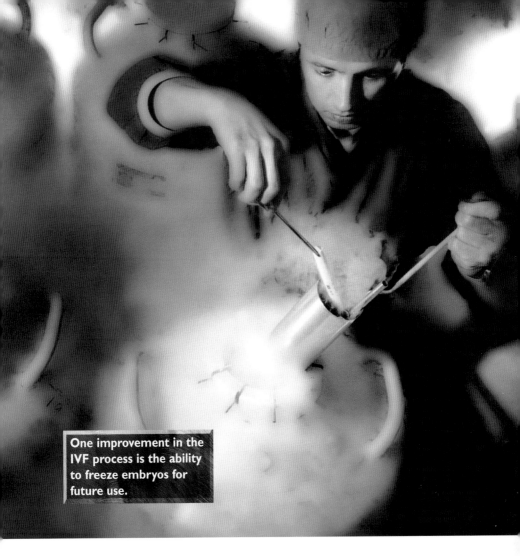

One improvement in the IVF process is the ability to freeze embryos for future use.

One improvement that is now being used keeps the fertilized embryo in the laboratory for a longer period of time (five to seven days versus twenty four hours), which gives it a chance to mature and develop more before it is placed in the woman's uterus. This also gives the mother's uterine lining more time to prepare for implantation. Embryos that can survive to reach this more advanced stage of growth appear to be hardier, which also increases the chance of success. Other experts are looking into ways to increase the success rate, to bring down the costs, and to lower the risks of multiple births or birth defects.

Despite the remarkable successes that have been achieved with IVF, the process does have its critics. Some people have expressed concern over just where the procedure could eventually lead. For example, when a woman has a number of fertilized eggs, she may opt to have some of them frozen. In August 2002, Helen Perry, the first British woman to be implanted with one of her own frozen eggs, gave birth to a daughter. The egg had been removed months earlier, frozen, stored, and then thawed before use. Perry said, "I think that egg freezing may come to be seen as the ultimate kind of family planning." These frozen eggs are kept in liquid nitrogen. U.S. ethical guidelines allow them to remain frozen and available for use for the woman's entire reproductive life. If she freezes her eggs, a woman can use them if her first IVF attempt fails or if she simply wants to get pregnant again in the future. The freezing of eggs, however, can lead to some unusual predicaments. For example, if a couple gets divorced, it would not be clear which partner would legally own the embryos. The law also does not state what should happen if both parents die. It remains to be decided whether the embryos would be destroyed or if it would be acceptable for a family member to claim them.

Women who want to focus on their careers may freeze their eggs and have them fertilized, then implanted, later in life.

Other questions also come to mind. For instance, some observers wonder whether infertile women alone should have the right to freeze their eggs. *Perhaps fertile women, who want to delay motherhood to start careers or for other reasons, should also be permitted to go through the IVF process, freeze their young eggs, and then return ten to fifteen years later and have those eggs implanted after the rest of their bodies have aged.* Already, there is a clinic in Los Angeles where fertile women can have their eggs collected, frozen, and saved. If healthy women can freeze their eggs for the future, some argue, then women who must have their ovaries removed for health reasons or women who may lose their ability to have children because of chemotherapy treatments for cancer should also be allowed to freeze their eggs.

Beyond the question of freezing eggs, there is also controversy over who should be able to use the IVF procedure itself. For example, some people question whether terminally ill patients who would not be able to conceive naturally should be allowed to use IVF. By the middle of 2002, more than 500 babies had been born to HIV-positive patients through the IVF process in Europe. Many people wonder whether people with the deadly HIV virus or any other fatal health condition should be permitted to become parents.

The ability to unite the egg and sperm outside of the human body is an incredible achievement—and one that also leads to other amazing concepts. If a woman cannot produce a viable egg, for instance, a donated egg from another woman can be used instead. If a woman can produce an egg, but her uterus will not be able to support a child for nine months, her egg can be carried by a surrogate mother instead. Although IVF is a miracle to childless couples who wish to have a baby, it can also lead to some tough questions and potential disasters.

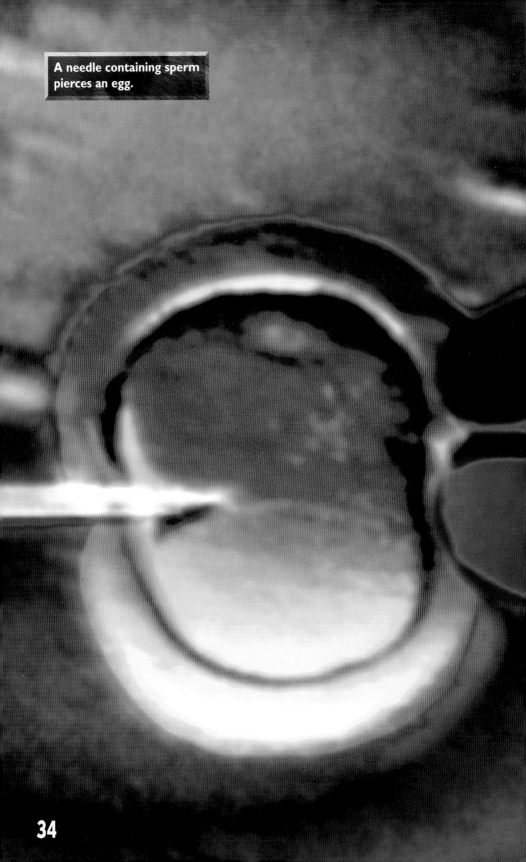

A needle containing sperm pierces an egg.

THE STORY OF DIANE BLOOD

English couple Diane and Stephen Blood had planned to have children ever since they were married. In February 1995, however, the unexpected happened. Stephen came down with bacterial meningitis. After he fell into a coma, he died. Just before his death, Diane convinced his physicians to take some of Stephen's sperm and freeze it so that she would have the option to have his child after he was gone.

Using IVF, Diane Blood was able to become pregnant. She gave birth to Liam Stephen in 2002.

Once again, plans did not go as Diane had hoped. The Human Fertilisation and Embryology Authority (HFEA) in England, a group of twenty one British citizens who make most decisions related to IVF, said no when Diane attempted to use the frozen sperm. Because Stephen had been in a coma and could not give written consent for the use of his sperm, the HFEA ruled that Diane did not have legal access to it.

A three-year court battle followed with much media attention, but in the end, the HFEA's decision was overturned. Diane was allowed to take her husband's sperm from the hospital. She went to Belgium for the IVF procedure, and on December 11, 1998, a son named Liam Stephen Blood was born.

In February 2002, Diane returned to Belgium and had the IVF procedure done once more. She now has a second son, Joel Michael—a child born almost eight years after his father had died.

OTHER ISSUES

When fertilization happens in a laboratory or clinic rather than within the human body, the door is open for mistakes to be made. Though such instances are rare, errors certainly do occur. In Holland, in 1993, the egg of a woman named Wilma Stuart was accidentally fertilized by the wrong man's sperm. The mistake was not discovered until nine months later, when she gave birth to black twins, despite the fact that she and her husband were white. In New York, in 1998, a woman was implanted with the wrong fertilized egg. Donna Fasano, a white woman, gave birth to a black baby—and was then

Donna Fasano (center) filed a lawsuit against the clinic that implanted her with the wrong egg.

forced to give the child back to its biological parents. In October 2002, an IVF mistake was made at a London hospital. Two women had the wrong embryos implanted. The error was caught, but both women had to undergo medical procedures and take medication to prevent their possible pregnancies.

Along with the confusion and heartache in situations such as these come many questions. Of course, one pressing question is what happened and how the mistake was made. Perhaps most importantly, though, there is the more delicate matter of determining who the babies belong to—the birth parents or the biological parents. Although security measures are in place in all fertility clinics, human error has been enough to cause monumental problems. All of these incidents of IVF mistakes have forced ongoing updates in the security and organization of clinics. Future plans include repeated checks to make sure the right sperm, egg, and embryos go to the right people.

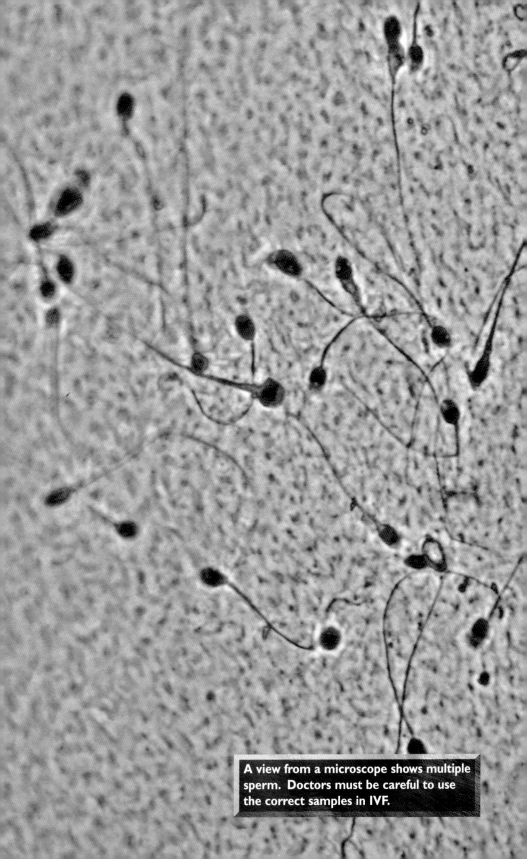

A view from a microscope shows multiple sperm. Doctors must be careful to use the correct samples in IVF.

TODAY IVF, TOMORROW A CLONE?

One of the doors that the concept of IVF has opened is a link to cloning, or the creation of genetic duplicates of people through the use of their specific DNA patterns. It has already been done successfully on several kinds of animals.

The science of cloning is a very controversial one, and there are valid arguments on each side.

One possible use for cloning is to save

Each person's DNA pattern is unique.

endangered animals. Many species, such as the giant panda, do not breed well in captivity. Cloning might be a way to keep the species alive. The first such experiment was with a guar, an Asian ox-like animal that had been on the edge of extinction. The guar embryo was implanted into a cow named Bessie. It was born in January 2002, but died of an illness called dysentery two days after its birth.

A project in Australia is even trying to bring back an extinct species. The thylacine, or Tasmanian tiger, is a marsupial that has been extinct for about seventy five years. A specimen of a baby thylacine, with most of its DNA intact, was found in a museum. Scientists are working to piece the genetic coding together again to see if they can produce another tiger. They are also looking at the possibility of cloning an extinct Spanish mountain goat called a burcado, and in the future, may consider cloning gorillas, ocelots, and pandas.

The cloning of humans has been the subject of many science-fiction books and movies. For Dr. Severino Antinori, an Italian embryologist who is actively involved in IVF, however, cloning is a real possibility. Antinori announced in 1998 that he planned to help infertile couples have children with cloning technology. He intended to use his fertility clinic in Rome to clone the male member of the couples to help them have a baby. He would inject the genetic material from the father directly into a woman's egg and then implant it through the IVF process. He has faced some staunch opposition. Many scientists and physicians believe that human cloning is far too dangerous and runs the risk of terrible birth defects. Many people also feel that it is not an ethical or moral thing to do. Despite these worries, Antinori has responded, "I want to bring society with me, and persuade people that it [cloning] is right in rare cases to help infertile couples." More than 1,500 couples so far have already volunteered to be part of Antinori's cloning trials.

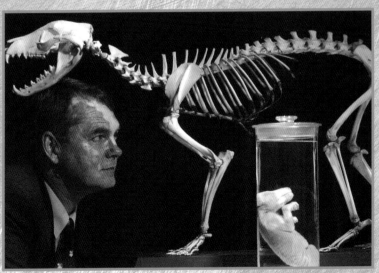

Many scientists want to use cloning to reproduce extinct animals.

Procedures like IVF have raised concerns about genetically designing babies.

DESIGNER BABIES

One of the biggest concerns about the future of IVF is the concept of designer babies. IVF doctors have already found a way to destroy embryos with certain health conditions, such as Down's syndrome, but some people question whether it is morally right to do so. It is also unclear who has the right to make that decision. Some researchers worry that, as it becomes easier to alter the genetic material of embryos, some parents will start to request—or reject— certain traits. Perhaps they will want only blond children or children with green eyes. Maybe they will want to make sure that

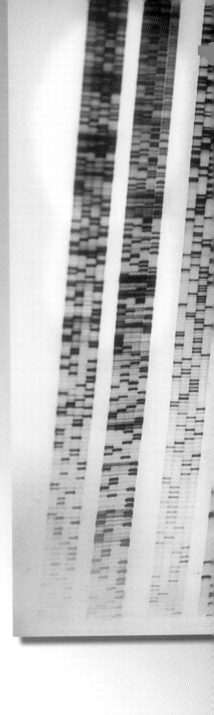

Many people question if requesting certain traits for babies should be allowed.

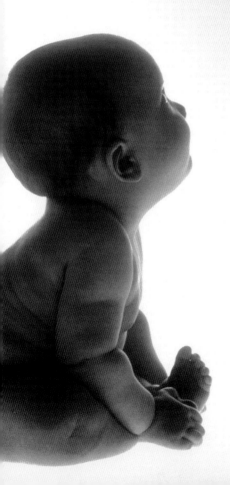

they have beautiful children. Some parents may only want a boy or a girl. In 2000, for example, a couple in Scotland lost their three-year-old daughter and wanted to use IVF specifically to have another daughter. Many people question whether requests such as these should be fulfilled.

Designer babies have already been produced for a few people. These children were genetically designed for one specific reason—to cure their older siblings of a life-threatening disease. The first such child was Adam Nash. Born in 2000, Adam was conceived to help cure his older sister through a bone marrow transplant. She was dying from a condition called Fanconi anameia, a genetic disorder in which the body is unable to make bone marrow.

In 2002, the case of Raj and Shahana Hashmi illustrated the controversy caused by the idea of designer babies. When the Hashmis' son Zain developed a potentially fatal blood condition called thalassemia, they found out that the cure could only be found in the umbilical cord cells of a sibling who was free of the disease and genetically compatible with Zain. They became eager to create an embryo that would match Zain's needs, and intended to discard any embryos that did not. The question raised among scientists and other observers of a case like this becomes one of whether the desire to harvest spare parts needed to save a living child is a valid reason to create another child. According to a ruling from the Human Fertilisation and Embryology Authority in England, the answer is yes. After a three-year battle, in October 2002, the authority stated that the Hashmis would be permitted to use "pre-implantation genetic diagnosis" to design a baby with the cells needed to cure their older son. In other words, the couple

A child stands on a walkway dedicated to IVF births outside a fertility clinic in Norfolk, Virginia.

would be able to go through the IVF procedure and select a particular embryo for implantation that would be genetically compatible with their son Zain. When asked about the ethics of what they were doing, Shahana Hashmi said, "This baby is going to be a special gift from nature, not a designer baby. We are not destroying anything. We are not hurting anybody."

It is up to organizations such as the Ethics Advisory Board of the U.S. Department of Health, Education and Welfare; the American Society for Reproductive Medicine; the American College of Obstetricians and Gynecologists; and the Judicial Council of the American Medical Association to answer the tough questions raised by cases such as the Hashmis'. It is the responsibility of these organizations to establish the guidelines and standards for the entire field, often lumped under the title ART, or artificial reproduction techniques. This is quite a challenging job.

Even with the many worries and risks, costs and complications, the miracle of test tube babies born through the process of in vitro fertilization is still a dream come true for thousands of families all over the globe. As the technology advances, questions and dilemmas will also continue to rise. Despite it all, people's desire to become parents will always remain strong, and IVF will provide a way for many to do just that.

GLOSSARY

AIDS acquired immune deficiency syndrome; a condition in which the body is unable to fight disease.

ART assisted reproductive technology.

capacitation a process through which sperm must go in order to be able to penetrate and fertilize an egg.

catheter a hollow, flexible tube inserted into the body to allow the passage of fluids (most often for urine in the bladder).

cloning the process of making a genetic duplicate based on a DNA pattern.

embryo in humans, the product that results from conception and implantation until the eighth week of growth.

endocrinologist a physician who studies the glands and hormones of the body and their related disorders.

endometriosis a condition characterized by endometrial tissue (which lines the uterus) that grows outside of the uterus and often causes pain and infertility.

follicle a small bodily cavity or sac, such as those that surround the eggs.

gynecologist a physician who specializes in the diagnosis and treatment of disorders that affect the female reproductive organs.

HIV human immunodeficiency virus; the virus that causes AIDS.

hyperstimulation overstimulation of a body system, such as hormonal medications cause in the body's natural hormone system.

infertility the inability to get pregnant or carry a baby to term after attempting to do so for the course of a full year.

in vitro Latin for "in an artificial environment".

ISCI intracytoplasmic sperm injection.

IVF in vitro fertilization.

laparoscopy a surgical procedure that uses a laparoscope—a slender, tubular instrument inserted through an incision in order to view pelvic cavities.

meningitis inflammation of the lining of the brain and spinal cord, usually caused by a bacterial or viral infection.

ovulation the production of ova or the discharge of eggs from the ovary.

polyspermia condition in which more than one sperm fertilizes a single egg.

pronucleus (plural pronuclei) the single-cell nucleus of a sperm or egg before the two nuclei fuse during fertilization.

speculum an instrument used to dilate the opening of a body cavity (such as the vagina and cervix) for medical examination.

STDs sexually transmitted diseases, such as chlamydia and gonorrhea.

ultrasound the use of ultrasonic waves for diagnostic or therapeutic purposes, such as to monitor a developing fetus or to see an image of an internal body structure.

vasectomy the surgical removal of all or part of the male's vas deferens (the main duct through which semen is carried), usually as a means of sterilization.

FOR FURTHER INFORMATION

Organizations
The Center for Applied Reproductive Science
Johnson City Medical Center Office Building
408 State of Franklin Rd.
Johnson City, TN 37604
Phone: (423) 461-8880

The International Council on Infertility Information Dissemination, Inc.
P.O. Box 6836
Arlington, VA 22206
Phone: (703) 379-9178

The Society for In Vitro Biology
9315 Large Dr. West, Suite 255
Largo, MD 20774
Phone: (301) 324-5054

ABOUT THE AUTHOR

Tamra B. Orr is a full-time freelance writer and author. She has written more than two dozen nonfiction books for children and families, including *Fire Ants, The Journey of Lewis and Clark, The Biography of Astronaut Alan Shepard* and *The Parent's Guide to Homeschooling.* Orr attended Ball State University and received a B.S. degree in secondary education and English in 1982. Orr lives in Portland, Oregon, with her husband and four children, who range in age from seven to eighteen. She enjoys her job as an author because it teaches her something new every day.

INDEX